Robson da Silva Teixeira

Interoperability between information resources

Robson da Silva Teixeira

Interoperability between information resources

For access to scientific production

ScienciaScripts

Cover image: www.ingimage.com

This book is a translation from the original published under ISBN 978-3-330-76326-5.

Publisher:
Sciencia Scripts
is a trademark of
Dodo Books Indian Ocean Ltd. and OmniScriptum S.R.L publishing group

120 High Road, East Finchley, London, N2 9ED, United Kingdom
Str. Armeneasca 28/1, office 1, Chisinau MD-2012, Republic of Moldova, Europe
Printed at: see last page
ISBN: 978-620-8-08883-5

INTEROPERABILITY BETWEEN INFORMATION RESOURCES TO ACCESS SCIENTIFIC

ROBSON DA SILVA TEIXEIRA

PhD candidate in Museology and Heritage at the Federal University of the State of Rio de Janeiro (UNIRIO). He is currently head librarian at the Library of the Institute of Physics of the Federal University of Rio de Janeiro (UFRJ).

Abstract: With the exponential growth of information resources on the Internet, Libraries are faced with the great challenge of managing their physical and electronic resources in an integrated manner. The aim of this article is therefore to design a virtual space, with a single interface for accessing scientific production, which allows interoperability in the information search phase in the online catalog and the Virtual Health Library. This interface will provide access to a variety of traditional types of content, such as books, journals, technical reports, software, as well as complex multimedia entities that mix text, images, videos and data.

Keywords: Interoperability; Technological Standards; Computer Systems Information; Scientific Production.

SUMMARY

1 INTRODUCTION

There is a huge field of work to be explored in organizing the scientific output generated at Brazilian public universities. Of course, it is up to these institutions, which produce and disseminate knowledge, to look after these documents, since they are essential for the construction, knowledge and development of History and Research in the country. It is clear that the production of digital information is currently far greater than the capacity to guarantee access to it, and that there are no consolidated strategies to guarantee long-term access to information of continuing value. Currently, researchers at the UFRJ Institute of Physics have two data providers to make available and access the scientific output generated in the community: the Aleph/Minerva System and the VHL (Virtual Health Library). However, there is no single interface that encompasses these two data providers, which brings together open access files in the same environment, i.e. which concentrates searches for academic works in a single interface, are forced to carry out more than one search, because the information is scattered in the Minerva database (UFRJ) and in the institution's databases (BVS-UFRJ). This situation leads to a waste of time that could be avoided if users could concentrate their searches for academic works on a single institutional interface. This would give users more time to dedicate to analyzing the bibliography retrieved and to carrying out the research itself. The literature consulted in this area shows that it is possible for data providers with fields and metadata from different technological solutions to become interoperable.

For access to this information to be effectively viable, the system in which it is stored must be able to generate processes that are interoperable with other systems. A truly interoperable organization is able to maximize the value and reuse potential of the information under its control, while also being able to effectively exchange this information with other equally interoperable organizations, allowing new knowledge to be generated from the identification of relationships between previously unrelated data sets. The level of interoperability will be referred to as the degree of commitment or coupling between systems (institutions, digital libraries) to make them interoperable and a measure of the effort to do so. The Open Archives Initiative (OAI) presents itself as an option to answer the question of making the different data providers at UFRJ interoperable. The methodology used for this is bibliographic research and documentary research. The use of bibliographic research allowed for the identification and qualitative analysis of the contributions of the main authors to the construction of the conceptual framework elaborated on the object of study. Documentary research is very close to bibliographic research and also allowed for the identification and access to specialized documentation on the information resources analyzed.

In today's contemporary and highly competitive world, environmental variables exert constant pressure on institutions. In this scenario, organizations must constantly evaluate their services in order to make information-based decisions. There is therefore a huge field of work for management in public university libraries - and this work is urgent. Of course, it is up to universities, as producers and disseminators of knowledge, to take care of their historical documentation and, above all, to ensure excellence in the provision of services and products to their users.

In this sense, it is important to note that:

It is the role of libraries, understood in this article as Information Units, to offer specialized services and products, adding value to them, with creativity in their implementation and format without losing focus on users and the satisfaction of their information needs. (ANJOS *et al.,* 2012, p. 90)

According to Levacov (1997): "Technology emerges as a catalyst for change, particularly important and poignant for libraries, since it creates new needs and alters old paradigms established over many centuries." The advent of new Information and Communication Technologies (ICTs) and the constant changes in technological processes and innovations are causing libraries to rethink the act of producing, accessing and distributing information, seeking to meet the new perspectives of meeting the needs of their users, locally or remotely. With the emergence of information and communication technologies (ICTs), new demands have arisen, creating a changing universe with new challenges for so-called information professionals. This professional is responsible for managing information and knowledge. The new paradigm increases the weight "of information and knowledge in the added value of products and the importance of the quality of human resources and their education in order to increase competitiveness" (BUARQUE, apud BARRETO, 2006).

The new world order has created new demands in the training of professionals and in information management. Information must be retrieved and shared in order to generate knowledge that can be managed and marketed. According to Lambert (2000), the role of the information professional who is "in tune" with technological developments and the changes that have taken place in access to information, which make life easier for the user, will always be based on the use of new technologies to meet the information needs of researchers and any type of user.

As the information professional responsible for the Library of the UFRJ Institute of Physics, I feel obliged to bring the Library, with its services and products, ever closer to the users who use it, so that information reaches users in a faster, more practical and effective way.

The motivation for this research project came from reflecting on the importance of building a single interface for searching information in the Aleph/Minerva System and the VHL (virtual health library). At the moment, users of the Physics Institute Library are forced to carry out more than one search, because the information is scattered in the Minerva database (UFRJ) and in the institution's databases (VHL-UFRJ).

This situation leads to a waste of time that could be avoided if users could concentrate their searches for academic works on a single institutional interface. They could thus have more time to devote to analyzing the bibliography retrieved and to carrying out the research itself.

Starting an information dissemination project in a virtual environment that doesn't yet have any standards and then being forced to move towards standardization requires a lot of work and can happen in a very painful and embarrassing way, either because you didn't think it would be important one day, or because the urgency didn't allow it to be implemented. It could happen that when users go to use the system, there may already be other systems that they want to integrate with the one they are using. Then there's a problem: does the information system they're using "talk" to an open standard? If not, it will probably be doomed to failure, or, at best, it will be used by people who are not interested in contributing to its development, probably using it in a very restricted way. Although the existing systems at UFRJ are currently standardized, more in-depth research is needed to answer the main question of this research. It is believed that this measure will contribute to faster and more efficient dissemination of the scientific information contained in these information systems.

It should be emphasized that this project aims to contribute to giving institutional visibility to the Library of the UFRJ Institute of Physics and represents a significant part of the public research effort in Medical Physics in Brazil. Last but not least, it should also foster institutional strategies for adhering to the Open Access initiative for scientific knowledge in the country.

In view of the above, the aim of this research was to design a virtual space, with a single access interface, which allows interoperability in the information search phase in the Aleph/Minerva system and in the UFRJ VHL (Virtual Health Library). From this perspective, three specific objectives were defined:

1. Identify the state of the art on Interoperability issues between databases and information systems
2. Identify the main interoperability standards and protocols applied to electronic information resources;
3. To identify the possibilities of interoperability between two information resources made available by the Physics Institute Library.

At the end of the work, a single interface was designed that included fields and metadata from the different technological solutions available at the Library of the Institute of Physics of the Federal University of Rio de Janeiro (UFRJ). It should be emphasized that this research aims to contribute to giving institutional visibility to the Library of the Institute of Physics of UFRJ and represents a significant part of the public research effort in Medical Physics in Brazil.

Digital space is believed to be capable of storing a variety of traditional types of content - books, periodicals, technical reports, software - as well as complex multimedia entities that mix text, images, video and data.

It is necessary to make managers aware of the urgency and importance of guaranteeing long-term access to the information generated by the Institute's researchers, so that the library seeks to carry out work linked to the interests of the scientific community, where it participates, questions and discovers values.

2 THEORETICAL FRAMEWORK

The theoretical framework was developed in two parts: the first part gives an account of the environment in which the research will be carried out, i.e. a brief history of the Institute of Physics of the Federal University of Rio de Janeiro (IF/UFRJ), the PKnio Susseking Rocha Library, the IF/UFRJ Library *website*, the Medical Physics course, the Aleph/Minerva information system and the UFRJ VHL (Virtual Health Library). The second part presents a preliminary bibliographical review of Usability, Interoperability and Technological Standards for Interoperability.

3 THE INSTITUTE OF PHYSICS AT THE FEDERAL UNIVERSITY OF RIO DE JANEIRO: A REAL SPACE IN THE HISTORY OF PHYSICS IN BRAZIL

The IF/UFRJ was created on March 19, 1964, on the occasion of the university reform, which brought together the physics courses then existing in schools and colleges in Rio de Janeiro belonging to the University of Brazil, and is part of the Center for Mathematical and Nature Sciences (CCMN) of that university.

Before it was created, the Physics course was part of the National Faculty of Philosophy (FNFi) and brought together five Physics departments, with the aim of training bachelors and graduate teachers.

After its foundation, with the hiring of professors, the Institute was set up to implement research activities and prepare for postgraduate studies, which had not existed until then.

4 THE PLNIO SUSSEKIND ROCHA LIBRARY

According to Brandao and Carvalho (2009) the Information Units obey the rules established by the Library System of the Federal University of Rio de Janeiro (SiBI/UFRJ), which aims to support teaching, research and extension programs, developing in accordance with the institution's planning and forming its collection in line with the syllabus of the different subjects offered at undergraduate and postgraduate level.

SiBI/UFRJ makes the library's collection and services available through the Aleph/Minerva system, developed for processing, storing and retrieving bibliographic and multimedia information.

As part of SiBI/UFRJ, the Plinio Sussekind Rocha Library, linked to the Institute of Physics, has a collection of approximately 13,000 books and 267 periodical titles (national and foreign).

5 THE IF/UFRJ LIBRARY *WEBSITE*

The *website of* the Plinio Sussekind Rocha Library of the IF/UFRJ, which uses the *Wordpress.org* platform, was created at the end of 2012 with the intention of facilitating the process of searching for information, as well as providing a collaborative space for information and knowledge, covering a range of services and products aimed at the academic community of Physics and related areas.

Since the IF/UFRJ is a unit with a consolidated research tradition, the benefits that interested parties seek were taken into account when designing the *site*, such as saving time in obtaining data or, according to Arellano (2001, p. 12), "useful and relevant information via specialized information sources [...]". In this respect, Information Units are being designed to fulfill this function by facilitating simple and effective access to *online* resources.

The Federal University of Rio de Janeiro (UFRJ) is a teaching, research and extension institution, based in the city of Rio de Janeiro and divided into four centers: Center for Mathematical and Nature Sciences (CCMN), Center for Technology (CT), Center for Health Sciences (CCS), Center for Letters and Arts (CLA), Institute of Philosophy and Social Sciences (IFCS) and Center for Legal and Economic Sciences (CCJE). The Institute of Physics is part of the CCMN and was created by CONSUNI Resolution 22 of March 19, 1964, and is made up of five departments: Medical Physics, Mathematical Physics, Nuclear Physics, Solid Physics and Theoretical Physics. The institute's research and postgraduate activities began in the early 1970s and its researchers have played an active role in the national and international scientific community. The Plinio Sussekind Rocha Library, linked to the Institute of Physics, is a sector library with a collection of approximately 10,000 books and 250 periodicals (national and foreign). Its aim is to promote access to information by providing support for teaching, research and extension activities so that the university can fulfill its role as a global trainer of individuals.

6 MEDICAL PHYSICS COURSE AT UFRJ

All over the world, a great deal of effort is currently being put into what has come to be known as Medical Physics. The enormous development in the area of diagnostic instruments and therapies in medicine increasingly requires the participation of professionals from other areas, especially physicists, in activities involving ionizing radiation. In this universe, Rio de Janeiro's position is privileged, because in addition to UFRJ's institutional capacity with various Institutes and Postgraduate Programs involved in some way with Medical Physics, other institutions have joined together in an innovative project and formulated the Qualification in Medical Physics, based and coordinated by UFRJ's Institute of Physics; planned and executed by a council made up of representatives from each of the participating institutions. The course is professional and should provide the minimum requirements for practicing the profession of Medical Physicist, giving students the conditions to continue their training. The various institutions involved in the student's training must act in a co-responsible, interdisciplinary and concomitant manner with regard to the student. The UFRJ institutions are: the Institute of Physics (IF), the Institute of Biomedical Sciences (ICB), the Alberto Luiz Coimbra Institute for Postgraduate Studies and Engineering Research (COPPE) and the Clementino Fraga Filho University Hospital (HUCFF). The institutions outside UFRJ are: the Radioprotection and Dosimetry Institute (IRD), the Nuclear Energy Institute (IEN), both of which are subordinate to the National Nuclear Energy Commission (CNEN), and the National Cancer Institute (INCA).

7 UFRJ LIBRARY AUTOMATION PROCESS

The Federal University of Rio de Janeiro (UFRJ) is a teaching, research and extension institution, based in the city of Rio de Janeiro and divided into four centers: Center for Mathematical and Nature Sciences (CCMN), Center for Technology (CT), Center for Health Sciences (CCS), Center for Letters and Arts (CLA), Institute of Philosophy and Social Sciences (IFCS) and Center for Legal and Economic Sciences (CCJE). The Institute of Physics is part of the CCMN and consists of five departments: Medical Physics, Mathematical Physics, Nuclear Physics, Solid Physics and Theoretical Physics. According to Brandao, it was only in 1993 that the automation process began with the registration of manuscripts and, in 1994, the entry of books and scores began. However, the methodology used was time-consuming, as the bibliographic information was described in spreadsheets and then entered into the University's Library Automation System. It was noticed that

The system's needs and requirements were enormous. Therefore, in 1996, with the support of the University, SIBI - Sistema Integrado de Bibliotecas (Integrated Library System), looked for software to manage library services, allowing the integration of collections and making them available through local and remote access to the academic community.

After an exhaustive analysis of service management systems, in 1997 we opted for the ALEPH software which, once implemented, gave rise to Base Minerva (www.minerva.ufrj.br), a virtual collective catalogue for remote consultation, which is used to record and disseminate the content of important collections from other UFRJ information units, such as archives, specialized collections and museums.

8 MAIN CHARACTERISTICS OF ALEPH

According to the manufacturer, "ALEPH 500™ is an integrated system for automating libraries and research centers", developed by the company Ex Libris and acquired by UFRJ for this purpose in 1997. Its main features are:

Flexibility and adaptability - it can be used both by individual libraries and by library systems or consortia. It also has customization components with parameters to accommodate the requirements of institutions of all types and sizes.

Ease of use - user-friendly workflow and intuitive graphical interfaces.

Openness - based on international industry standards such as open URL, XML, OAI, LDAP, ISO ILL, and RFID, which guarantees its operability with other systems and durability, as the software project evolves continuously.

The system has the following integrated modules:

WEB OPAC - which is the public catalog that serves as a customer portal and has a completely customized interface, according to institutional needs.

ADAM - digital resource management module, works in conjunction with cataloging module

Cataloging - integrates the cataloging function with all the other functions of the system

Acquisition and periodicals - controls the processes of acquiring bibliographic material, including the management of orders, documentation and delivery.

Circulation - controls collection circulation processes

BSE - controls interlibrary loan requests

9 VIRTUAL HEALTH LIBRARY (BVS) - UFRJ

The UFRJ Virtual Health Library is simulated in a virtual Internet space formed by a network of health information sources. Users can interact and navigate in the space of one or more information sources, regardless of their physical location. Information resources include AIP (publications of the American Institute of Physics), OVID / MEDLINE, Springer, Scielo, among others. According to Martins (2006), among the information resources made available by the VHL-UFRJ is the Capes periodicals portal, which allows the academic community, made up of 70 federal higher education institutions, to have electronic access to the full content of national and international periodicals, through a multidisciplinary portal on the Internet. The Portal has become one of the main mechanisms for keeping the academic community of the Federal University of Rio de Janeiro (UFRJ) up to date with national and international scientific production. Students, teachers and researchers can access, download, copy and print, in part or in full, publications from the world's most renowned research centers. The Portal also brings together other databases that can be accessed free of charge via the *web* or subscribed to by CAPES.

According to Martins (2006), among the information resources made available by the BVS-UFRJ is the Capes periodicals portal, which allows the academic community, made up of 70 federal higher education institutions, to have electronic access to the full content of national and international periodicals through a multidisciplinary Internet portal. The Portal is not just a public database, but a set of licenses acquired from foreign companies so that a certain number of institutions can access the electronic journals and databases contracted. The Portal has become one of the main mechanisms for keeping the Brazilian academic community up to date with national and international scientific production. It can be accessed from any terminal connected to the Internet, from a participating institution. Students, teachers and researchers can access, download, copy and print, in part or in full, publications from the world's most renowned research centers.

Once the empirical framework has been presented, the theoretical framework will follow, which will deal with Usability, Interoperability and Technological Standards for Interoperability, since understanding these issues is essential for thinking about the interoperability of systems. The aim was to investigate in the literature the main factors that contribute to user satisfaction with this type of system.

According to Santos (2006), interfaces developed without meeting usability requirements lead to poor performance and a reduction in the quality of the user's interaction with an application. The level of usability of an interface is improved by considering user satisfaction, in an approach that values the experience of using the product.

According to Santos (2006), the usability of human-computer interaction has been an area of growing interest for Brazilian researchers. However, designers and developers of computerized systems make little use of ergonomic parameters in the design and testing of usability and interface evaluations; much more consideration is given to user satisfaction with the interface used. For the author, usability is a term used to define the ease with which people can use a tool or object in order to perform a specific and important task. Usability can also refer to methods of measuring usability and the study of the principles behind the perceived effectiveness of an object. According to Santos (2006) usability usually refers to the simplicity, ease, efficiency and uniformity with which an interface, computer program or website can be used. At this point, a link is made between usability and interoperability, because for Campos (2007) interoperability also refers to the ability to transfer and use information between systems efficiently and uniformly, requiring standardization and flexibility at a certain level. It relates to integration, cooperation, exchange, interaction and working together.

From the perspective of building a virtual space, it is also necessary to present the issue of interoperability and its importance when searching for information in two or more information retrieval systems.

According to Souza Netto (2008), since the emergence of the Internet and the consequent increase in electronic information sources, the term interoperability has been used to support discussions about the various levels of management and development of these resources. Thus, the concern to ensure that an organization's systems, procedures and culture are managed in such a way as to enable the exchange and reuse of information has become vital.

Interoperability in a hybrid information environment requires the adoption of norms and standards at all levels. These libraries must support functions in a similar way, or at least adhere to certain standards, so that information can be exchanged. Interoperability in this context aims to make coherent services available to users from technically different components managed by different organizations.

The Online Dictionary for Library and Information Science (ODLIS) defines the term interoperability as:

"The ability of a hardware or software system to communicate and work effectively in the exchange of data with another system, usually of a different type, designed and produced by a different supplier." (Online..., 2004).

According to Sayao (2008), interoperability is the ability of a system to communicate transparently with another system. For a system to be considered interoperable, it is very important that it works with open standards or ontologies, whether it is a portal system, an educational system or even an e-commerce system, nowadays we are increasingly moving towards the creation of standards for systems. In the field of information technology, interoperability is the exchange of information and/or data via computers. Interoperability is also the ability to communicate, to execute programs through various functional units, using common languages and protocols. For Marcondes (2002), interoperability has several facets, but technical interoperability is perhaps the one responsible for keeping information systems interoperable, but there are other important concepts, such as semantic interoperability. According to the author, semantic interoperability is related to the meaning or semantics of information originating from different resources and is solved by the adoption of common or mappable information representation tools, such as metadata schemes, classifications, thesauruses and, more recently, ontologies. An example of a question addressed by this facet of interoperability might be the following: what does "author" mean for an information resource? Is it the same thing as "creator" for another resource?

The challenge of designing coherent services for a diversity of users from components that are technically different and managed by different organizations requires a sophisticated degree of cooperation that can be differentiated (ARMS, 2000; ARMS et al., 2002). Interoperability is the way in which machines and systems communicate, and the ability of different systems (e.g. Digital Library Management Systems), through standards, agreements or protocols, to operate together in order to perform a task, but this system implies the orderly exchange of content, so there has to be a standardization of the metadata that makes up the files. It has to be a single web interface, allowing several digital libraries to be consulted simultaneously. However, it is important to note that digital collections require well-structured metadata schemes to describe digital objects at various levels: administrative, structural and descriptive, because along with the digital object, it must be surrounded by structured information.

With regard to technological standards for interoperability, it is known that integrating the various sources of information under a single user interface is considered extremely important in the process of making resources available. These should be based on open international standards, whenever possible, and the architecture should be flexible, allowing for the development of systems that support new implementations in the medium and long term, among other things. The adoption of standards aims to organize information in a structured way in a context of such heterogeneous information sources.

According to Sayao (2008), making information resources fully interoperable is no easy task, as it requires a high level of technical, content and organizational interoperability. Knowledge of the main communication standards and protocols, such as Z39.50, OAI PMH, SRW/U; metadata for representing information, such as MARC, Dublin Core and referential links, such as the OpenURL standard, technologies of interest to the study in question, are standards that have contributed significantly to the search for resource integration. According to Marcondes (2001), the Open Archives Initiative (OAI) *is an* effective, efficient, economical and viable technical solution for scientific communities to rebuild scientific communication practices and processes, manage cooperative management systems and bibliographic control mechanisms, preserve memories and consolidate their *corpus of* knowledge. Based on Marcondes (2002), it can be said that the Open Archives Initiative (OAI) aims to create technological mechanisms to make different repositories interoperable, working according to the Open Archives proposal.

It should be noted that, although there is a level of difficulty, there are records of successful experiences in the area of interoperability, or mapping of items that should be taken into account.

In a proposal developed as part of her master's thesis, Oliveira (2005) presents the possibilities of making heterogeneous databases interoperable, so that they can be searched through a single interface. The researcher used the LILACS, MEDLINE and ACERVOS ONLINE FIOCRUZ databases as a research tool. She analyzed a number of facets such as: the description procedures and structures of the databases, looking for similarities and differences in the indexes and display fields of the databases; thematic representation, characterized by the use of the DeCS and MeSH Thesaurus; and the technologies used and/or compatible with them. However, it should be noted that in proposing interoperability, the author restricted herself to databases and online collections. The aim of this project is to create a single interface for the information resources of the Institute of Physics, which would also involve the issue of usability.

In another master's thesis, Souza Neto (2008) set out to investigate the methodological and technological alternatives applied to electronic resources, with a view to identifying interoperability standards and protocols and their possible integration with the library's physical resources through integrated access. However, the researcher has focused her studies on the context of a hybrid library and has not addressed the issue of creating a single interface for searching for information. In this project, the intention is to develop a methodology for interoperability between the systems described in such a way as to make it possible to search the two existing systems from a single interface.

10 INTEROPERABILITY BETWEEN INFORMATION RESOURCES

According to Souza Netto (2008), since the emergence of the Internet and the consequent increase in electronic information sources, the term interoperability has been used to support discussions about the various levels of management and development of these resources. Thus, the concern to ensure that an organization's systems, procedures and culture are managed in such a way as to enable the exchange and reuse of information has become vital.

Interoperability in a hybrid information environment requires the adoption of norms and standards at all levels, in which both electronic and paper-based information sources are used. The focus of the concept is on services, which adapt to the new digital context in an effort to

transformation and reorganization of the traditional library (TAMMARO; SALARELLI, 2008, p. 118). These libraries must support functions in a similar way or at least adhere to certain standards so that information can be exchanged. Interoperability in this context aims to make coherent services available to users from technically different components managed by different organizations.

According to Sayao (2008), interoperability is the ability of a system to communicate transparently with another system. For a system to be considered interoperable, it is very important that it works with open standards, whether it is a portal system, an educational system or even an e-commerce system, nowadays we are increasingly moving towards the creation of standards for systems. In the field of information technology, interoperability is the exchange of information and/or data through computers, as well as the ability to communicate, execute programs through various functional units, using common languages and protocols.

For Marcondes (2002), interoperability has several facets, but technical interoperability is responsible for keeping information systems interoperable, but there are other important concepts, such as semantic interoperability. According to the author, semantic interoperability is related to the meaning or semantics of information originating from different resources and is solved by adopting common or mappable information representation tools, such as metadata schemes, classifications, thesauruses and, more recently, ontologies. An example of a question addressed by this facet of interoperability might be the following: what does "author" mean for an information resource? Is it the same thing as "creator" for another resource?

The challenge of designing coherent services for a diversity of users from components that are technically different and managed by different organizations requires a sophisticated degree of cooperation that can be differentiated (ARMS, 2000; ARMS et al., 2002).

With regard to technological standards for interoperability, it is known that integrating the various sources of information under a single user interface is considered extremely important in the process of making resources available. These should be based on open international standards, wherever possible, and the architecture should be flexible, allowing for the development of systems that support new implementations in the medium and long term. The adoption of standards aims to organize information in a structured way in a context of such heterogeneous information sources.

According to Sayao (2008), for information resources to be fully interoperable, the task is not easy, as it requires a high level of technical, content and organizational interoperability. Knowledge of the main communication standards and protocols, such as Z39.50, OAI-PMH, SRW/U; metadata for representing information, such as MARC, Dublin Core and referential links, such as the OpenURL standard, technologies of interest to the study in question, are standards that have contributed significantly to the search for resource integration.

The advent of Information and Communication Technologies (ICTs) and the constant changes in technological processes are making libraries rethink the act of producing, accessing and managing information, seeking to meet the new perspectives of meeting the needs of their users, locally or remotely. New demands have arisen, creating a changing universe with new challenges for information professionals, who are responsible for managing information and knowledge, the added value of products and the importance of the quality of human resources and their education in order to increase competitiveness" (BUARQUE, apud BARRETO, 2006). According to Levacov (1997): "Technology emerges as a catalyst for change, particularly important and poignant for libraries, since it creates new needs and alters old paradigms established over many centuries".

The new world order has created new demands in professional training and information management. Information must be retrieved and shared in order to generate knowledge that can be managed and marketed. According to Lambert (2000), the role of the information professional who is "in tune" with technological developments and the changes that have taken place in access to information, which make life easier for the user, will always be based on the use of new technologies to meet the information needs of researchers and any other type of user. The professionals responsible for managing university libraries must bring the library, with its services and products, ever closer to the users who use it, so that information reaches users in a faster, more practical and effective way.

According to Marcondes (2001), the Open Archives Initiative (OAI) *is an* effective, agile, economical and viable technical solution for scientific communities to rebuild scientific communication practices and processes, manage cooperative management systems and bibliographic control mechanisms, preserve memories, thus promoting the consolidation of their *corpus of* knowledge. Based on Marcondes (2002), it can be said that the OAI aims to create technological mechanisms to make different repositories interoperable, working according to the Open Archives proposal.

It should be noted that, although there is a level of difficulty, there have been successful experiences in the area of interoperability, or mapping of items that should be taken into consideration. In a proposal developed as part of her master's thesis, Oliveira (2005) presents the possibilities of making heterogeneous databases interoperable, so that they can be searched through a single interface. The researcher used the LILACS, MEDLINE and ACERVOS ONLINE FIOCRUZ databases as a research tool. She analyzed some facets such as: the description procedures and structures of the databases, looking for similarities and divergences in the indexes and display fields of the DeCS and MeSH Controlled Vocabularies; and technologies used and/or compatible. However, it should be noted that in proposing interoperability, the author restricted herself to databases and online collections.

11 METHODOLOGICAL PROCEDURES

For this project, bibliographic research and documentary research were carried out. According to Gongalves (2003, p. 35), bibliographic research is characterized by the identification and qualitative analysis of the contributions of various authors on a given subject. Adopting this type of research allows for the identification and qualitative analysis of the contributions of the main authors in order to build a conceptual framework about the objects of study.

Documentary research is very close to bibliographical research, according to Gil (1995, p. 73): "The essential difference between the two lies in the nature of the sources". While bibliographic research basically uses the contributions of various authors on a given subject, documentary research uses materials that have not yet received an analytical treatment, or that can still be re-elaborated according to the objects of the research.

In order to meet the proposed objectives, the methodology used will be described below, broken down into 3 stages:

In the first stage, the state of the art on Interoperability between databases and information systems was drawn up.

In the second stage, an analytical reading of the texts was carried out in order to obtain a theoretical basis for the proposed themes. Some points discussed in the literature about the nature and function of the Open Archives Initiative (OAI), which aims to create technological mechanisms to make different repositories interoperable, were also read and used as an answer to the question: How can data providers with fields and metadata from different technological solutions be made interoperable?

In the third stage, documentary research was used in order to describe and identify the main standards and protocols aimed at interoperability applied to the electronic information resources made available by the Physics Institute Library, since documentary research allowed the identification of and access to specialized documentation on the information resources analyzed in this work (Aleph/Minerva System and VHL-UFRJ).

In parallel to these steps, the fields of the Minerva and VHL databases were mapped, in order to make the correspondence between the two systems. In the light of this framework, the multiple search interface was designed, within the appropriate guidelines and consistent with the desirable usability standard for an information system, allowing the Aleph/Minerva System and the VHL-UFRJ to be consulted simultaneously. With this measure, the library seeks to carry out work linked to the interests of the scientific community, where it participates, questions and discovers values.

12 RESULTS

The project is in the implementation phase and the study carried out has made it possible to affirm that it is possible to develop a virtual space capable of providing access to a variety of traditional types of content - books, periodicals, technical reports, *software* - as well as complex multimedia entities that mix text, images, video and data, using the Open Archives Initiative. The technological standard and communication protocol used was Z39.50, as it is based on open international standards, allowing the development of systems that support new implementations in the medium and long term, and the metadata used to represent the information was MARC. All these technological standards and communication protocols allow for a single interface that includes fields and metadata from so many different technological solutions in the Library of the Institute of Physics of the Federal University of Rio de Janeiro (UFRJ).

13 FINAL CONSIDERATIONS

The main objective of this research - to design a virtual space with a single access interface that allows interoperability in the information search phase in the Aleph/Minerva system and in the UFRJ VHL (Virtual Health Library) - has been achieved. The interface is in the final testing phase and is expected to be available to users in the first half of 2017.

As a reflection for future research, it is believed that there are strong indications that this work should be continued, there is a need for constant revisions and updates so that the interface is always in convergence with the needs of the users who use it.

It is believed that the interface fulfills the role of disseminating information, enabling the Information Unit to carry out work that is linked to the interests of the scientific community, as well as saving the user time by optimizing the service.

Considering that storage does not mean availability when needed and is a cost that is generally underestimated when creating systems that aim to be effective, it is thought that one of the competitive advantages of the virtual reference service is precisely the fact that virtual information sources can be updated quickly, require less manpower and do not need physical space for storage.

As this is an experience report, the implementation of a digital service has made it possible to meet the demands of the users of the Information Unit more quickly and effectively, and to offer an efficient and easily accessible response, since the contemporary world demands efficient management from organizations, which can be facilitated with the support of intelligent resources offered by technology and the various information systems available.

It is hoped that by publishing this report, other initiatives will be implemented, corrected and/or updated, thus enabling the Library to be closer to its users and offer quality services and products with unlimited access.

BUDGET

Hiring an Individual:

Web Developer Analyst to create a single interface for the systems Aleph/Minerva and BVS .

Quantity: 1

Individual Monthly Value: R$2,000.00

Annual Total for Individuals: R$24,000.00

Permanent Material:

- Intel Core 2 Duo Computer with 2GB RAM, 200GB HD, Monitor 17" LCD

Quantity: 1

Individual price: R$1,600.00

- 1.3KVA No Break

Quantity: 1

Individual price: R$400,00

Total Permanent Material: R$2,000.00

Consumables (paper, toner): R$500,00

Total budget for the project: R$26,000.00

REFERENCES CONSULTED

ARMS, W.Y. *Thoughts about interoperability in the NSDL: draft for discussions.* August 2000. Available at: <http:// www.cs.cornell.edu/wya/papers/NSDL-Interop.doc>. Accessed on: March 15, 2011.

. A spectrum of interoperability: the site for science prototype for the NDSL. *D-Lib Magazine,* Reston, Virg., v. 8, n. 1, jan. 2002.

BARRETO, Angela Maria. Information Management: tool for production or meaning? *Inf. & Soc.: Est.,* v. 16, n. 2, p. 54-68, jul/dez 2006.

BRANDAO, Dolores Castorino. *Biblioteca Alberto Nepomuceno da Escola de Musica da UFRJ* : do raro ao virtual. Rio de Janeiro: Editora UFRJ, 1993. 120 p.

CHIZZOTTI, Antonio. *Pesquisa em cidências humanas e sociais.* 2. ed. Sao Paulo: Cortez, 1995.

FONSECA, Nadia Lobo da. *About books, memory and identity:* a reading of the initial years of Physics and the physicists of UERJ. 2009.147 f. Dissertation (Master's in Social Memory). Postgraduate Program in Social Memory, Federal University of the State of Rio de Janeiro, Rio de Janeiro, 2009.

GIL, Antonio Carlos. *Como elaborar projetos de pesquisa.3.ed.* Sao Paulo: Atlas, 1995. 159 p.

GONQALVES, Elisa Pereira. *Conversations on beginning scientific research.*

Campinas: Ed. Alinea, 2003. 80 p.

HILLWAY, T. *Introduction to Research,* 1964. pp. 256-58

LAMBERT, M. B. M.A. The new role of the information professional in the information society. *Cidncia da Informagao,* Brasilia, v. 35, n. 3, p. 125-135, Sept/Nov. 2000.

LEVACOV, Marilia. Virtual libraries: (r)evolution? *Cidncia da Informagao,* Brasilia, v. 26, n. 2, p. 125-135, May/Aug. 1997.

MARCONDES, Carlos Henrique; SAYAO, Luis Fernando. Integration and interoperability in access to electronic information resources in S&T: the proposal of the Brazilian Digital Library. *Cidncia da Informagao,* Brasilia, v. 30, n. 3, p. 24-33, Sep./Dec. 2001.

Digital documents and new forms of cooperation between S&T information systems. *Cidncia da Informagao,* Brasilia, v. 31, n. 3, p. 42-54, Sep./Dec. 2002.

MARTINS, Maria de Fatima Moreira. *Study of the use of the CAPES Portal in the process of knowledge generation by Biomedical researchers:* applying the critical incident technique. 2006. 128 f. Dissertation (Master's Degree in Information Science) - Fluminense Federal University, Rio de Janeiro, 2006.

OLIVEIRA, Viviane Santos de. *Searching for interoperability between different databases:* The case of the Fernandes Figueira Institute Library. 2005. 116 f. Dissertation (Master's in Health Information and Communication Management) - Sergio Arouca National School of Public Health - FIOCRUZ, Rio de Janeiro, 2005.

RAYWARD, W. Boyd.Introduction. In: RAYWARD, Warden Boyd (ed.) *International organization and dissemination of Knowledge.* Selected Essays of Paul Otlet.Amsterdam, New York, Oxford, Tokio: Elsevier, 1990. p.3.

SAYAO, L.F.; MARCONDES, C.H. The challenge of interoperability and new perspectives for digital libraries. *Translnformagao,* Campinas, 20(2): 133-148, May/August, 2008.

SOUZA NETTO, Erica de. *Integrated access to information resources: focus on interoperability*. 2008. 132 f. Dissertation (Master's in Information Science) - Fluminense Federal University, Rio de Janeiro, 2008.

TAMMARO, Anna M.; SALARELLI, A. *The digital library.* Brasilia: Briquet de Lemos, 2008.

UFRJhist6ria.Dispon'welem:<http://www.ufrj.br/pr/conteudo pr.php?sigla=HISTORIA>. Accessed on: March 13, 2011.

Printed by Books on Demand GmbH, Norderstedt / Germany